Andy visits a fire house.
He talks to firefighters.

Firefighters
wear thick coats.
They also wear
rubber boots.
The coat and boots
keep them safe.

5

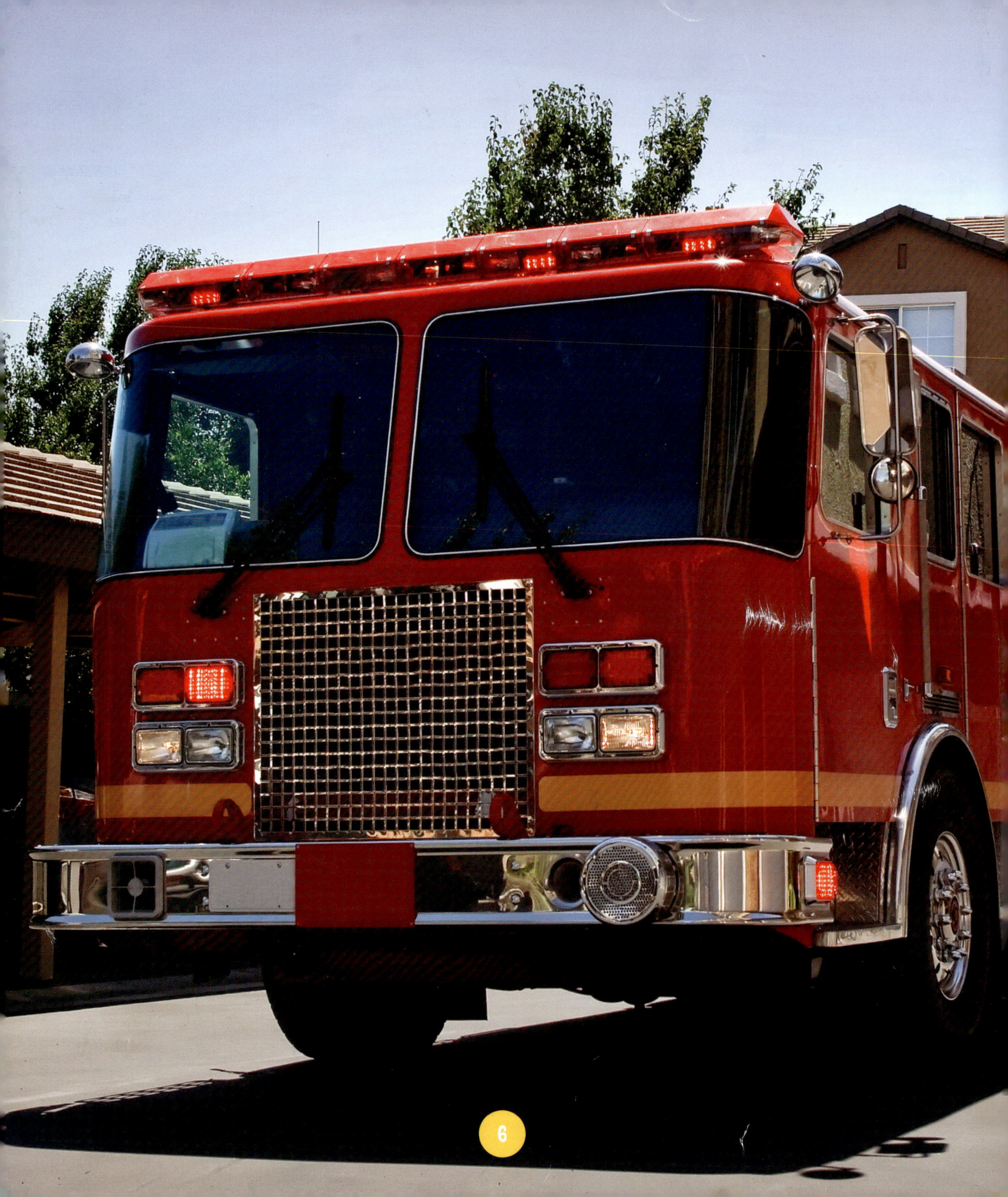

They drive trucks.
The trucks
have red lights
and a siren.

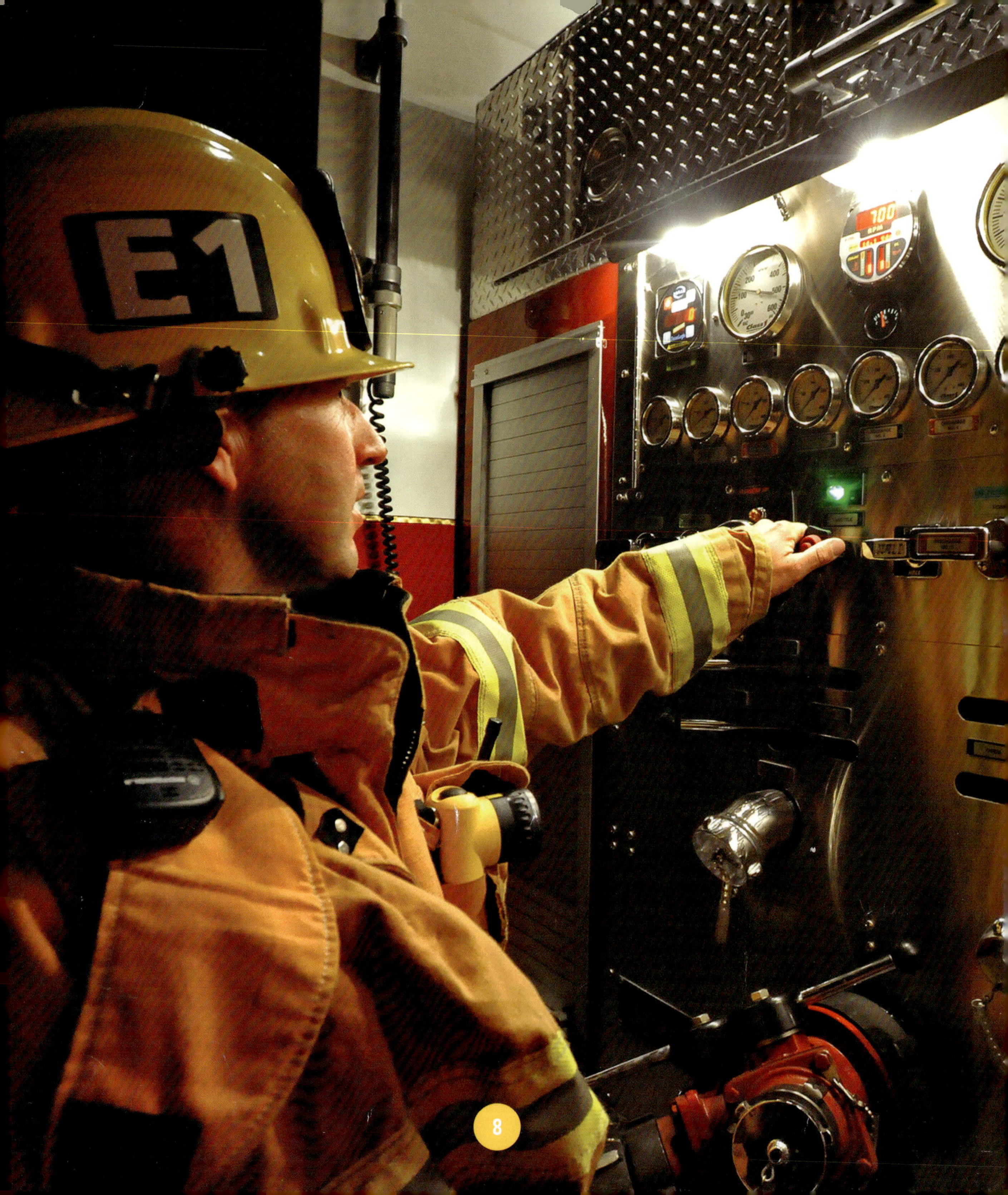
E1

The trucks hold tools to fight fires. They have ladders. They have long hoses, too.

10

The fire bell rings.
Time to go!

12

The firefighter faces the flames. He sprays water on the fire.

Firefighters saved the day!

15

SEEN AT A FIRE HOUSE

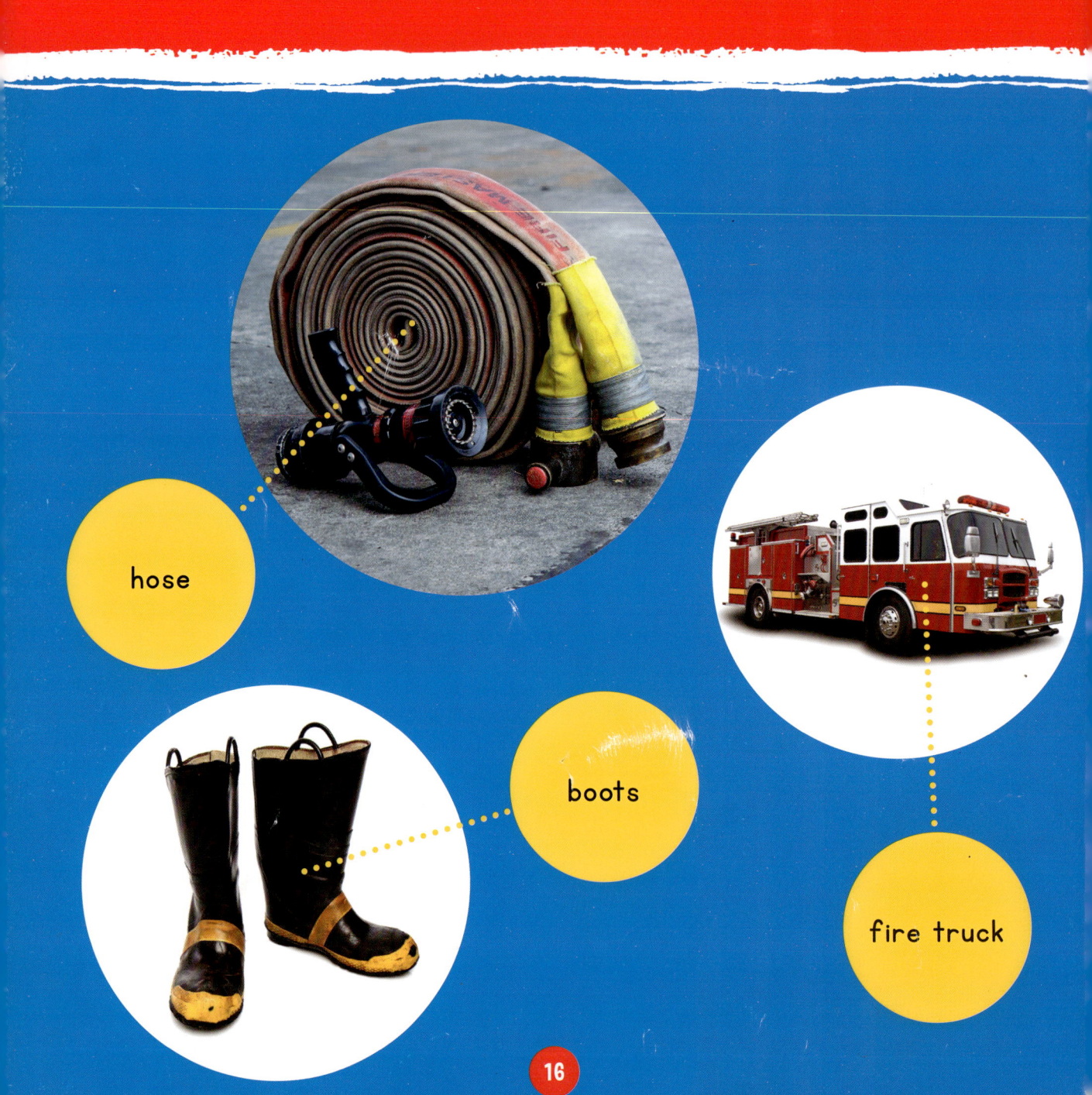

hose

boots

fire truck